AF718573

DYEMAKING WITH AUSTRALIAN FLORA

RIGBY LIMITED • ADELAIDE • SYDNEY
MELBOURNE • BRISBANE • PERTH

First published 1974
Reprinted August 1974
Copyright © 1974 Handweavers and Spinners Guild of Victoria
Library of Congress Catalog Card Number 72-91964
National Library of Australia Card Number
and ISBN 0 85179 664 8
All rights reserved

Wholly designed and set up in Australia
Printed in Hong Kong

DYEMAKING with Australian FLORA

The Handweavers and Spinners
Guild of Victoria

RIGBY

This book was produced for the Handweavers and Spinners Guild of Victoria by an Editorial Committee consisting of:

Convenor

VALDA MARTIN

Committee Members

JEANNE LAMPRELL
SHIRLEY LUCAS
VAL MARKS

CONTENTS

Section of a floor rug, woven from fleece coloured with natural dyes

INTRODUCTION

By 1788, the year of the first British settlement in Australia, chemical dyes were already widely used in Europe, and vegetable dyes were gradually losing their popularity. The European settlers brought woven and dyed fabrics with them to their new country, and consequently they had no need to explore for vegetable dyes. They were far more interested in discovering the agricultural and mineral potential of Australia than in extracting colour from its native vegetation.

The Aborigines who inhabited the continent at this time depended on plants for sustenance, and so vegetable dyes were not important to them. When they wanted colour for their tribal paintings or to decorate their bodies for ceremonial purposes, they did not consider the juices of fruit, flowers, roots, or bark, but used instead the pigments of the earth.

The famous nineteenth century British botanist, Sir Joseph Dalton Hooker, wrote: "Australian flora contains more genera and species peculiar to its own areas, and fewer plants belonging to other parts of the world than any other country of equal extent."

The plant dyes described in this book are based only on Australian native vegetation, but not every species has been included. This book is written as a guide and textbook, and it is the hope of the compilers that it will lead readers to discover dye colour from other plants at a time when people throughout the world are enjoying a revival of many ancient crafts, including spinning, weaving, and dyeing.

Most plants have dye colour. Roots, bark, leaves, flowers, and berries all yield colour. By mixing mordants and overdyeing, many shades from the one plant can be obtained. Different soils, climatic conditions, and time of year also have an effect on the colour produced.

The preparation of this book has been a fascinating project involving members of the Handweavers and Spinners Guild of Victoria and other interested people throughout Australia. They contributed samples and detailed records of their experiments, and the dyers themselves were responsible for selecting the plant material and mordants. Alum has proved to be the most popular of the mordants chosen, closely followed by copper, tin, iron, and chrome. Various combinations of these and other mordants have been tried successfully when dyeing with the Eucalypts and Acacias. Alum is readily available from local pharmacies.

From the outset dyers were asked to observe the laws governing the protection of native plants. In cases where plants were not "garden-grown natives," they were gathered from private properties with the permission of the owner. In these days of rapid development it is often possible to gather dyestuff simply by following in the wake of earth-moving equipment. After a storm, foliage of tall trees can be collected in order to make dyebaths. Many species of Australian plants can be cultivated in the home garden.

Wool has been used for the tests, because it is available throughout Australia and is receptive to dye. Fleece from Corriedale, Merino, Lincoln, English Leicester, Cheviot, and Border Leicester sheep, and crossbreds have been used. When samples of coarse and fine wools were dyed in the same dyebath, the finer consistently resulted in a paler shade. It should also be noted that fibres other than wool may be successfully dyed with Australian native flora.

The original record sheets of the various tests, with their samples of dyed wool attached, have been mounted in a book which remains the property of the Handweavers and Spinners Guild of Victoria. It may be inspected in the Guild Library.

PREPARING THE WOOL

Generally there are three processes involved in dyeing—scouring, mordanting, and dyeing. If you hurry through these procedures, you will not be successful.

Wool must be completely free of all grease and dirt to make it receptive to both mordant and dye. As much as fifty per cent of the weight of raw wool may consist of impurities of one sort or another. Wool also contains natural grease (yolk) and dried perspiration (suint), and usually the finer wools contain a higher percentage of natural impurities than the coarser wools. These impurities must be removed before processing can be carried out. To achieve this, soak the spun wool tied loosely in skeins, or the loose fleece, in cold water overnight. Rinse *several* times in cold water. *Do not wring, rub, or twist the wool at any time.*

Scour by repeated immersions in water hotter than the hand can bear for any length of time, using soap or a detergent prepared for use with wool. The wool could become harsh if a stronger detergent more suitable for cotton is used. Several rinses in warm water are essential, and it may be necessary to repeat the whole procedure several times. In areas where water is "hard," it is advisable to use rain water. Hard water can be softened by the addition of a little acetic acid or vinegar.

The *damp* wool is now ready for mordanting or, alternatively, for drying.

MORDANTING WOOL

Mordanting refers to the addition of a chemical for the purpose of fixing the colour in the fibre. The wool can be adversely affected if not kept completely submerged and covered during the whole process or if the temperature is suddenly changed. Enamel, glass, or stainless steel vessels are recommended because iron or aluminium will affect the mordanting. Uneven mordanting through insufficient mixing will cause uneven dyeing. If water evaporates during mordanting, remove the wool, replace lost amount with boiling water, stir, and then return the wool to the bath.

SOME MORDANTING AGENTS

Care should be taken in handling these chemicals, and they must be kept out of the reach of children. The amounts given are for one pound of dry, scoured wool but they may be reduced proportionately.

ALUM — POTASSIUM ALUMINIUM SULPHATE

85·05 grams (3 ounces) alum
28·35 grams (1 ounce) cream of tartar
0·45 kilogram (1 pound) wool
18·20 litres (4 gallons) water

IRON — FERROUS SULPHATE
14·18 grams ($\frac{1}{2}$ ounce) iron crystals
28·35 grams (1 ounce) cream of tartar
0·45 kilogram (1 pound) wool
18·20–20·48 litres (4–4$\frac{1}{2}$ gallons) water

CHROME — POTASSIUM DI-CHROMATE
7·09–14·18 grams ($\frac{1}{4}$–$\frac{1}{2}$ ounce) chrome
0·45 kilogram (1 pound) wool
18·20–20·48 litres (4–4$\frac{1}{2}$ gallons) water
Less chrome is used for fine wool.

COPPER — COPPER SULPHATE
7·09–14·18 grams ($\frac{1}{4}$–$\frac{1}{2}$ ounce) copper
0·45 kilogram (1 pound) wool
18·20–20·48 litres (4–4$\frac{1}{2}$ gallons) water

TIN — STANNOUS CHLORIDE
7·09–14·18 grams ($\frac{1}{4}$–$\frac{1}{2}$ ounce) tin crystals
56·70 grams (2 ounces) cream of tartar
0·45 kilogram (1 pound) wool
18·20–20·48 litres (4–4$\frac{1}{2}$ gallons) water

ACETIC ACID — It can be used as an addition to dyebaths, quantities $\frac{1}{2}$ teaspoon to 4·55 litres (1 gallon) in warm water.

OXALIC ACID — 7·09 grams ($\frac{1}{4}$ ounce) oxalic acid
0·45 kilogram (1 pound) wool
It is generally used as an additive with other mordants.

TANNIC ACID — 7·09 grams ($\frac{1}{4}$ ounce) tannic acid
0·45 kilogram (1 pound) wool
This is a strong astringent and may be used as an additive with other mordants.

METHOD FOR MORDANTING WOOL

ALUM — Dissolve the alum and cream of tartar with a small amount of water and add to 18·20 litres (4 gallons) of soft water. Immerse the *damp* wool, cover the container, heat gradually to boiling point, agitate, and let it simmer gently for one hour. If evaporation occurs, add boiling water so that the proportion of liquid to wool remains the same. Cool and allow the wool to stand overnight in the mordant. Squeeze it from this bath, roll loosely in a dry towel or place in a plastic bag, and then put in a cool, dark place.

Wool can be dyed immediately after mordanting, but it is better if kept in the damp condition for several days.

Rinse the mordanted wool well to remove surplus chemicals prior to putting it in the dyebath.

CHROME — Dissolve the chrome in 18·20–20·48 litres (4–4$\frac{1}{2}$ gallons) of cold water and follow the directions for mordanting wool with alum. It is *very important* to keep the wool completely submerged and the container covered while using chrome. Keep the mordanted wool in a dark place as chrome is very sensitive to light. Careful rinsing is necessary. Dyeing immediately after mordanting is recommended.

TIN — A small amount of tin added to a dyebath will help to brighten the colour. It is not generally used as a mordant but rather as an additive towards the end of dyeing.

When using tin, the wool does tend to become harsh and brittle. Wash it in soapy water after mordanting and rinse several times until the water is clear.

For mordanting, dissolve both tin and cream of tartar separately in cold water, and then add to the measured amount of cold water. Let it simmer gently for a few minutes. Allow to become cold, add *damp* wool, reheat gently, and then let it simmer again for twenty to thirty minutes. Rinse as above and use as required.

COPPER — Dissolve in cold water, let it simmer gently for a few minutes, cool, and then add *damp* wool and simmer for a further ten minutes. Cool and rinse the wool. Copper may also be used as an additive with other mordants towards the end of the dyebath.

IRON — When using iron, the wool is first dyed then mordanted. Let the wool simmer in the dyebath for approximately thirty minutes and then remove the wool while you add the iron and cream of tartar, dissolved in a little water. Mix the liquid well before returning the wool to simmer for ten to fifteen minutes and stir often. Rinse very thoroughly. It is advisable to keep a separate container for iron as it dulls the colour and affects subsequent dyes.

DYEING

Proportions of the basic quantities of the various materials needed to carry out the dyeing operation are given as follows:

Water	Dye material	Mordanted wool
0·57 litres (1 pint) water	28·35 grams (1 ounce) dye material	14·17 grams (½ ounce) mordanted wool
1·14 litres (2 pints) water	56·70 grams (2 ounces) dye material	28·35 grams (1 ounce) mordanted wool
2·28 litres (4 pints) water	113·40 grams (4 ounces) dye material	56·70 grams (2 ounces) mordanted wool
4·55 litres (1 gallon) water	0·23 kilogram (8 ounces) dye material	113·30 grams (4 ounces) mordanted wool
9·10 litres (2 gallons) water	0·45 kilogram (1 pound) dye material	0·23 kilogram (8 ounces) mordanted wool
18·20 litres (4 gallons) water	0·90 kilogram (2 pounds) dye material	0·45 kilogram (1 pound) mordanted wool

To prepare a stronger dye, use proportionately more dye material to the quantity of water. Variations in colour occur for many reasons. The locality in which the dye material is collected, seasonal conditions, the question as to whether the material is fresh or dry, and types of fleece, are but a few of them.

It is suggested that a record, similar to the following, be kept of all experiments. Attach the dyed sample to the record.

1. State whether the dye material is from foliage, bark, roots, flowers, et cetera. State whether plant is young or old.
2. Record the date the plant was gathered and the locality.
3. Name the type of wool if possible and the method of washing prior to dyebath.
4. Name the type of dyepot used, e.g. aluminium.
5. Name the mordant, if any, used, and describe your method of mordanting.

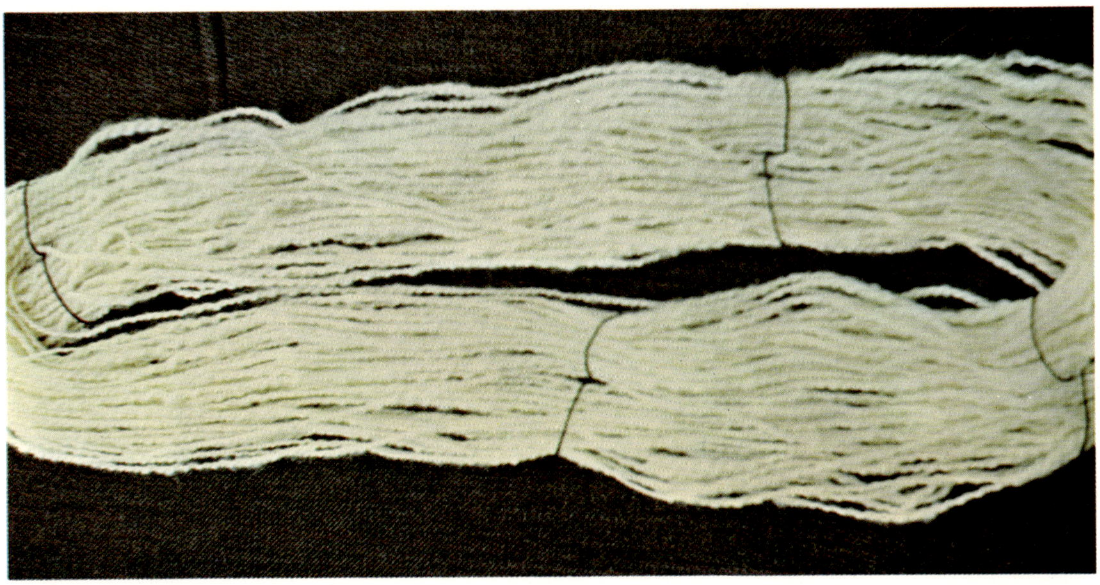

Several ties made through and around the skeins will prevent the wool from becoming tangled during washing, mordanting, and dyeing

BASIC METHOD

The *damp* wool may be added at the same time as the vegetation. Alternatively, strain the dye liquid, add the wool, then return the dyebath to boiling point, and let it simmer for a further thirty minutes. Stir back and forth frequently.

It is important to dampen the wool thoroughly, whether mordanted or not, before adding to any dyebath. Cover your vessel to prevent evaporation.

Cut up roughly the vegetation to be used, place it in the dyepot, cover with water, and let it simmer for one or two hours. Enamel, glass, or stainless steel vessels are recommended for use as they do not chemically alter the dye, like aluminium or iron dyepots.

Since some dye materials release their colour more readily than others, the length of boiling time varies accordingly. Occasionally the colour may be lost by over-boiling.

Rinse the dyed wool first in hot and then in cooler water. Keep rinsing until the water is clear. Squeeze out carefully and dry in the shade. Spun wool should be dried at slight tension. Alternatively, leave the wool to cool in the dyebath and then rinse it in cold water.

The above is a basic method for dyeing, but some vegetation requires quite different handling. For example, lichens, dried material, and bark are better if left soaking for a day or two prior to dyeing. Berries should be crushed. Long, slow simmering generally produces richer colours.

Satisfactory colours, without using any mordant, were derived from the leaves of the Boobialla, Kangaroo Paw, several Eucalypts and Mistletoes, the flowers of Sturt's Desert Pea, and the green fronds of the Staghorn Fern. However, only testing over a number of years will ascertain their colour fastness.

Test the colour fastness of your dyed wool before use. A simple method is to place a skein half in and half out of a box with its lid on. Leave in a sunny place for several days and then compare the two halves of the skein. Colours fade when subjected to bright light for long periods of time, and it is important to consider this fact if working on, for example, a wall hanging where pale shades are used in intricate patterns.

Skeins of wool should be dried under tension. Hang the skeins from one rod, and insert another through the bottom loops. Then exert slight tension on the yarn by hanging a bucket, filled with water, on each end of the lower rod

It is possible to obtain different shades from dyebaths. Keep adding wool at intervals until the dye is exhausted. Successive dyes will be progressively paler in colour. Red, orange, and then yellow dyes can result from the same dyebath made from the leaves of the Silver Leaf Stringybark.

Exciting colours may be obtained by OVERDYEING. Dye the wool first in one dyebath, rinse, and dry. Dampen again and then re-dye in a different dyebath. Three different experiments are listed:

Experiment 1 Dyeing with Native Cherry and alum mordant gave a good lemon-yellow to both crossbred handspun and commercially spun wool samples. The second dye, made from the bark and kino (resin-like substance) of the Pink-flowered Ironbark with iron, changed the colour of the crossbred handspun to a deep yellow and the commercial yarn to a good green.

Experiment 2 Original dyeing using the young leaves of the Blue Gum with alum produced a salmon pink in the crossbred wool used. When overdyed with the bark and kino of the Pink-flowered Ironbark and iron, the colour became orange-tan.

Experiment 3 Crossbred handspun was dyed to a deep yellow using the leaves and stems of the Forked Sundew. Overdyeing, with bark and kino of the Pink-flowered Ironbark and iron, produced a dull dark green.

COMPOUND MORDANTING

Compound mordanting can be effected in this way. Mordant with alum all wool to be used. Choose your dye subject and prepare a dyebath following the basic method. Add the wool, proceed with the dyeing, and then rinse thoroughly (Mark this Sample 1—alum mordant). To the same dyebath add a small quantity of chrome and a new portion of wool; dye and rinse this in a similar manner (Sample 2—compound mordant alum/chrome). To the same dyebath add a very small quantity of oxalic acid and a new portion of wool. Proceed with dyeing and rinsing as previously (Sample 3—compound mordant alum/chrome/oxalic acid).

Continued experimenting with compound mordanting will produce a range of colours from the one dye matter. Change away from the alum-mordanted wool and commence with wool mordanted in copper or chrome. Try your own combinations.

It is wise to keep a detailed record of all experiments as well as a small sample of the colour obtained.

Eucalyptus pauciflora, *Snow Gum, in the Australian Alps*

DYES FROM AUSTRALIAN NATIVE PLANTS AND TREES

At All Times Observe the Laws Governing the Protection of Native Plants.

Leaves, stems, roots, bark, berries, seeds, and flowers may be used separately or combined to produce dye colour.

All the samples are listed in alphabetical order of the botanical name. It is advisable to become familiar with these as there are many variations of the popular or common names. In cases where there are alternate botanical names (synonyms) for a particular plant, they are in brackets thus: (syn.—).

Leaves, stems, roots, bark, berries, seeds, and flowers may be combined or used separately to produce dye colour

Wattle blossom can be mordanted to produce many different shades of yellow and gold

Certain colours may be produced by two or more mordants (for example, alum and copper). This indicates that the mordants are combined in the one dyebath.

When distinctly different colours have been submitted from the same species, they are shown thus:

Mordant

alum—red, yellow, and gold

In hyphenated colours, the predominate colour is named first, for example, "gold-green" would mean a gold colour with green undertones.

A short story of the dye material is included where the plants or trees used to produce it are either involved with Australia's history and growth, or have some interesting aspect.

WATTLES

The early settlers gave the Australian Acacias the popular name of Wattle, derived from the Anglo-Saxon word "Watul" meaning a hurdle or shelter made from interwoven sticks. They were the first to use the young pliable growths of these and other native trees for making fences and wattle-and-daub houses. By "wattling" or weaving the saplings together, they formed the walls and daubed them with wet clay until it filled all the crevices.

Acacias should not be confused with Mimosas as they are a different genus. Although Australia has several hundred species of native Acacias, there are no Mimosas. Australian Wattles are cultivated in many countries for their floral beauty, and some especially for their bark which is used for tanning.

Ranging in size from low-growing shrubs in exposed, coastal areas to majestic trees in the moist mountain forests, Acacias are more widespread throughout Australia than Eucalypts. Some species thrive in the arid areas of Australia, where there is insufficient rainfall for Eucalypts.

The timber of several species of the Wattle is valuable. For example, the Blackwood, a tall tree of the mountains, has a tough timber with an attractive grain sometimes marked in the shape of a fiddleback. This is highly prized for furniture and cabinet making. Hickory is very similar in appearance to Blackwood. The timber is also tough and it is used for furniture and tool handles. The wood of the Mulga, from the arid inland of Australia, is used for small, highly-polished souvenir pieces. "Mulga" is an Aboriginal word meaning "a long narrow shield" made from the wood of this Acacia. It was also used for spear heads, and as a slow-burning wood it was ideal for camp fires. The Aborigines ground the seeds of Mulga into a flour from which "cakes" were made. They also chewed the twigs and roots of some Acacias for nourishment, and spun the remaining stringy, fibrous part into thread to make nets, girdles, and bags.

Wattles regenerate rapidly after a bushfire. In fact seeds which have lain dormant for more than half a century have been known to germinate following a fire. The heat cracks the hard outer coating of the seed and makes it ready for growth with the next rain. The leaves are sometimes eaten by stock in times of drought but not all species provide good fodder. Some are poisonous to animals.

The bark of the Black Wattle and of the Silver Wattle is used for tanning. Bark stripping is done in spring and early summer. The common name of the Silver Wattle refers to the silvery patches on the bark. The Australian Golden Wattle is considered to produce the most important tanning bark in the world. A native of South Australia, Victoria, and south-western New South Wales, it is used as Australia's floral emblem and is featured on the Coat-of-Arms.

The spectacular Cootamundra Wattle is probably the most widely cultivated, both in Australia and overseas, although it is natural only to a small area of New South Wales near the town of Cootamundra.

ACACIA ANEURA—Mulga
DRIED LEAVES AND TWIGS
Mordant
alum—fawn

ACACIA BAILEYANA—Cootamundra Wattle
LEAVES AND STEMS
Mordant
alum—lemon, yellow-gold, and gold
chrome—pink-tan, dark gold, and gold-green
copper—dark green
alum and copper—fawn-gold
alum and tin—gold
alum, tin, and tannic acid—orange
alum, tin, tannic acid, and acetic acid—orange-tan
alum and oxalic acid—tan-fawn
alum and acetic acid—fawn-gold
alum, cream of tartar, and copper—fawn-yellow
tin, cream of tartar, and oxalic acid—bright gold-yellow
FLOWERS, STEMS, AND LEAVES
Mordant
alum—yellow
FLOWERS
Mordant
alum—lemon-yellow
SEED PODS
Mordant
copper—fawn

ACACIA CULTRIFORMIS—Knife-leaf Wattle
LEAVES
Mordant
alum—yellow
FLOWERS
Mordant
alum—gold

ACACIA CUNNINGHAMI—Curracabah or Motherumbung
LEAVES
Mordant
alum—pale fawn
tin—yellow
copper—light brown
iron—grey
iron and copper—grey
FLOWERS
Mordant
alum—fawn
tin—gold
copper—brown
iron—green
iron and copper—grey

ACACIA DEALBATA—Silver Wattle
LEAVES
Mordant
alum—yellow-fawn
chrome and copper—green

ACACIA DECURRENS (syn. A. NORMALIS)—Queen Wattle or Early Black Wattle
LEAVES
Mordant
alum—yellow and lemon
copper—green
iron—green-grey

ACACIA ELATA—Cedar Wattle
LEAVES
Mordant
alum—gold
chrome—gold-green
FLOWERS
Mordant
chrome—dull gold

ACACIA HARPOPHYLLA—Brigalow
LEAVES
Mordant
alum—fawn
tin—gold
copper—brown
iron—grey

tin and copper—green
tin and iron—green

ACACIA IMPLEXA—Hickory or Lightwood

LEAVES

Mordant
alum—yellow
copper—brown
(no mordant—fawn)

ACACIA LONGIFOLIA—Sydney Golden Wattle or Sallow Wattle

LEAVES

Mordant
alum—lemon
chrome—dull yellow

ACACIA MEARNSI (syn. A. MOLLISSIMA)—Black Wattle or Green Wattle

LEAVES

Mordant
alum—gold
alum and tin—fawn

FLOWERS

Mordant
alum—gold

SEED PODS

Mordant
alum and tin—grey-fawn

BARK

Mordant
alum—fawn-tan

ACACIA MELANOXYLON—Blackwood or Mudgerabah

LEAVES

Mordant
alum—lemon-fawn

ACACIA PODALYRIAEFOLIA—Mount Morgan Wattle or Queensland Silver Wattle

FLOWERS

Mordant
alum—lemon-fawn and lemon
copper—green-fawn
chrome—yellow
iron—green-grey

SEED PODS

Mordant
alum—fawn

ACACIA PRAVISSIMA—Ovens Wattle or Tumut Wattle

LEAVES

Mordant
alum—fawn-brown, yellow-gold, and orange-fawn
chrome—fawn and gold-brown

ACACIA PROMINENS—Golden Rain Wattle or Gosford Wattle

LEAVES

Mordant
alum—gold

ACACIA PYCNANTHA—Australian Golden Wattle, Golden Wattle, or Broadleaf Wattle

LEAVES

Mordant
alum—gold

ACACIA RETINODES—Wirilda

LEAVES

Mordant
alum—dull yellow and fawn

ACACIA SALIGNA—Golden Wreath Wattle, Willow Wattle, or Weeping Golden Wattle

LEAVES

Mordant
alum—lemon

ACACIA SPECTABILIS—Mudgee Wattle or Pilliga

LEAVES

Mordant
alum—fawn
copper—fawn-gold
chrome—deep gold
iron—green-fawn

AGONIS FLEXUOSA—Willow-Myrtle, Western Australian Willow-Myrtle, or Peppermint Tree

LEAVES

Mordant
alum—grey-fawn, lemon, yellow, and gold
copper—green-fawn
chrome—fawn
iron—grey

BARK

Mordant

alum—fawn

(no mordant—pale fawn)

ANGOPHORA COSTATA (syn. A. LANCEOLATA)—Smooth-Barked Apple-Myrtle, Rusty Gum, or Smooth-Barked Apple

LEAVES

Mordant

alum—fawn, lemon, and yellow

tin—gold

copper—light brown

iron—grey

iron and copper—dark grey

BARK

Mordant

alum and acetic acid—lemon-fawn

alum, acetic acid, and oxalic acid—pale tan

alum, acetic acid, oxalic acid, and chrome—pale fawn-grey

alum, acetic acid, and chrome—pale fawn

alum, acetic acid, and copper—green-fawn

alum, acetic acid, copper, and iron—green-grey

KANGAROO-PAW

There are several varieties of Kangaroo-Paw, so called because of the flowers' resemblance to the paw of this animal. Native to Western Australia and cultivated in many other states, it is a perennial herb with flat, ribbon-shaped leaves springing directly from the underground stem. This is the floral emblem of Western Australia. The flowers vary with the different species. Some are green, yellow, or red; one is green and red.

ANIGOZANTHOS MANGLESI—Red-and-green Kangaroo-Paw

LEAVES

Mordant

alum—fawn-pink

copper—green-grey

(no mordant—pink)

FLOWERS

Mordant

alum and copper—green-fawn

ANIGOZANTHOS VIRIDIS—Green Kangaroo-Paw

LEAVES

Mordant

alum and cream of tartar—grey and grey-green

ANTHEROSPERMA MOSCHATUM—Southern Sassafras, Sassafras, or Tasmanian Sassafras

LEAVES

Mordant

alum—yellow and grey-fawn

ATALAYA HEMIGLAUCA—Whitewood

LEAVES

Mordant

alum—fawn

chrome—light brown

copper—green

A clump of the spectacular Red-and-green Kangaroo-Paw

When mordanted with alum, the leaves of the Coast Banksia produce a lemon-yellow dye

BANKSIA

This entirely Australian genus was named after Sir Joseph Banks, the eminent botanist who travelled with Captain Cook on his first voyage when he discovered eastern Australia in April 1770. During the short stay in Botany Bay, Banks and others of the ship's crew made short excursions inland. It was then that Banks collected the first samples of Australian flora.

Some of the taller species share the common name of Honeysuckle Tree. Bees and the various Honeyeater birds favour their flowers. Sometimes the nectar of Banksias ferments and intoxicates the bees. Aborigines made an alcoholic drink by placing the flowers in water for a few days until fermentation took place.

BANKSIA INTEGRIFOLIA—Coast Banksia or White Honeysuckle

LEAVES

Mordant

alum—lemon-yellow

BANKSIA MARGINATA—Silver Banksia

LEAVES

Mordant

alum—yellow

BAUHINIA CARRONI—Bauhinia

FLOWERS AND LEAVES

Mordant

alum—lemon
chrome—fawn-yellow
alum, tin, and cream of tartar—yellow
copper—dark grey
iron—grey

BILLARDIERA SCANDENS—Common Apple-Berry

FLOWERS, LEAVES, AND STEMS

Mordant

alum—lemon

BLECHNUM NUDUM—Fishbone Water-fern

LEAVES

Mordant

alum—green-yellow

BORONIA MEGASTIGMA—Brown Boronia or Scented Boronia

LEAVES

Mordant

alum—lemon
iron—gold
copper—yellow

KURRAJONG

The Kurrajong is a tree ranging from small to medium in size and is natural to the eastern states, although it has been extensively planted as a street tree in dry areas. The trunk is so soft and spongy that the timber has no commercial use, but the leaves provide valuable fodder for stock in times of drought.

The Aborigines used the lacy or net-like bark of the Kurrajong for their bags and nets. They relished the soft fleshy roots, which they boiled and ate.

It is sometimes known as the "rattle-trap" tree because of the loud rattling noise made by the hard, woody, seed pods when shaken by the wind. The early settlers made use of the bark, which they macerated in water to produce a lace-like bast to convert into ropes, cordage, and coarse paper.

The bell-shaped flowers hang in clusters, and the seed pods are black.

BRACHYCHITON POPULNEUS (syn. STERCULIA DIVERSIFOLIA)—Kurrajong

LEAVES

Mordant

alum—orange-gold

cream of tartar—pale fawn

alum and cream of tartar—lemon

BRACHYCHITON RUPESTRE—Queensland Bottle Tree

LEAVES

Mordant

alum—fawn

tin—green

copper—light green

iron—grey-green

BRACHYSEMA LANCEOLATUM—Swan River Pea

LEAVES

Mordant

tin, oxalic acid, and cream of tartar—lemon

BOTTLEBRUSH

Callistemon is a wholly Australian genus of about twelve shrubs with strikingly beautiful flowers. Their colours range from red and pink to pale yellow-green. As the common name suggests, they are shaped like a bottlebrush. Some of the Callistemons are valued for their timber which is hard and tough and is used for tool handles. The wood shavings can be woven or plaited. *Callistemon citrinus,* the Crimson Bottlebrush, is a native of Victoria, New South Wales, and Queensland. The leaves are lemon-scented when crushed.

CALLISTEMON CITRINUS (syn. C. LANCEOLATUS)—Crimson Bottlebrush, Lemon-scented Bottlebrush

LEAVES

Mordant

alum—yellow-fawn

iron—grey

chrome—dull gold

CALLITRIS COLUMELLARIS—White Cypress-Pine, Murray Pine, Bribie Island Cypress, or Coastal Cypress

LEAVES

Mordant

alum—gold

FRUIT

Mordant

alum—fawn

CALLITRIS CUPRESSIFORMIS (syn. C. RHOMBOIDEA)—Port Jackson Pine or Oyster Bay Cypress-Pine

LEAVES

Mordant

alum—fawn
tin—lemon
copper—green
iron—dark green

CALOPHYLLUM INOPHYLLUM—
Mangrove

LEAVES

Mordant
alum—fawn
tin—gold
copper—brown
iron—green

CAPPARIS MITCHELLI—Native Orange or Native Lime

LEAVES

Mordant
alum—bright yellow
chrome and potash—dull orange
tin and cream of tartar—yellow
copper—yellow-green
copper and tin—green
iron—green

CASUARINA CUNNINGHAMIANA
River Sheoke or Fire Oak

LEAVES

Mordant
alum—gold and lemon
copper—green-fawn
iron—grey

The Crimson or Lemon-scented Bottlebrush

Sturt's Desert Pea, one of Australia's largest wildflowers, is so named because the explorer, Charles Sturt, gathered it near Broken Hill in 1844

STURT'S DESERT PEA

There are two members of the Clianthus species. One, Sturt's Desert Pea, or Glory Pea, is native to Australia, while the other is native to New Zealand.

Sturt's Desert Pea grows naturally in Western Australia, Northern Territory, New South Wales, and in South Australia where it is the floral emblem. After rare rains, the dormant seeds burst to produce, with amazing rapidity, the great sprawling vines covered with red, pea-shaped flowers with dark red, almost black, shiny blotches at the centre of the flower.

Although the Desert Pea is generally associated with the explorer Sturt, it was William Dampier who first collected it in 1699. The original specimen is now kept in the Oxford University Herbarium.

CLIANTHUS FORMOSUS—Sturt's Desert Pea or Glory Pea

FLOWERS AND LEAVES

Mordant

alum—fawn-tan

copper—green-tan

alum and copper—gold-green

(no mordant—pink-fawn)

CORREA REFLEXA—Red Correa

LEAVES AND STEMS

Mordant

alum—yellow

copper—gold-brown

(no mordant—fawn-yellow)

CYATHEA AUSTRALIS—Tall Treefern

LEAVES

Mordant

alum—yellow

DAVIESIA LATIFOLIA—Hop Bitter-pea

LEAVES AND STEMS

Mordant

alum—fawn

DICHONDRA REPENS—Kidney Weed

WHOLE PLANTS

Mordant

alum—green-fawn

DODANAEA CUNEATA—Hop Bush

LEAVES AND STEMS

Mordant

alum and cream of tartar—lemon and yellow-fawn

FORKED SUNDEW

With the exception of north-western Victoria, this soft perennial herb grows in the eastern states and South Australia.

During the second World War when there was a shortage of ink, some consideration was given to growing this plant for the commercial extraction of ink, but this was found to be impracticable.

DROSERA BINATA—Forked Sundew

WHOLE PLANTS

Mordant

alum—dull gold

EPACRIS IMPRESSA—Native White-heath

FLOWERS, LEAVES, AND STEMS

Mordant

alum and cream of tartar—lemon-yellow

(no mordant—fawn)

EREMOPHILA MITCHELLI—Budda, Sandal Box, or Sandalwood

LEAVES

Mordant

alum—green-yellow

chrome—dark gold and gold

tin and cream of tartar—brown

copper—deep green

iron—brown

FLOWERS

Mordant

copper—dark brown

This Gum Tree at Wentworth, New South Wales, clearly displays the scar left on its trunk by Aborigines, who many years ago removed a long slab of bark in order to fashion a canoe

GUM TREES

The popular name, Gum Tree, was given to the Eucalypts by Governor Phillip, the founder of the British Colony in Australia. He was referring to the "kino" or resinous, gum-like substance which they secrete. In 1788 the Messmate-Stringybark became the first species to be scientifically described and named. The generic name, Eucalyptus, is derived from the Greek words, *eu* (well) and *kalyptos* (covered, in relation to the cap which tightly covers the stamens in the bud). The gums belong to the large family of Myrtaceae as do the Tea Trees *(Leptospermums)* and Bottlebrushes *(Melaleucas* and *Callistemons)*.

Eucalypts are native only to Australia, Papua New Guinea, and a few of the adjacent islands, although they have been grown successfully in a wide range of climates throughout the world. The American gold diggers, for instance, took with them some seeds of the Eucalypts when they returned to California from the Australian gold rush of the 1850s.

It was Dr White, Surgeon to the First Fleet, who first recognised the medicinal value of the Gum leaves in 1788 when he used oil distilled from the Peppermint Trees near Sydney Cove. Nearly half a century later commercial extraction of the oil commenced, following successful research carried out by Baron von Mueller, the first Government Botanist for Victoria.

The Aborigines had many different uses for the Eucalypts. They spun the bark fibres of some species into threads, used slabs of bark as shields, and painted pictures on the inner layer of the Stringybark. Their canoes were made from several species. The method of making a canoe was to cut a long thick section of bark from the trunk of a tree, heat it in ashes until it became pliable, and then bend it to a canoe shape. To complete the construction, they sealed the ends with clay.

In arid areas, water could be found in the roots of various Mallees. The Aborigines would dig these roots from the ground, cut them into pieces, and then stand them in "bowls" made from hollowed-out gnarls of Gum Trees. In this way as many as five pints of clear water could be obtained from one root. The Aborigines ate the powdered roots of Mallees and in some cases ground the seeds between stones after soaking them in water. These were either eaten raw or cooked in hot ashes. They obtained the nectar from the flowers of the smaller Gum Trees and particularly from the large-flowered Rose-of-the-West. They also gathered a sugary substance, made by scale insects, from the leaves and buds of some Gum Trees. This has the popular name of Manna. The knowledge of food and water supplies from these plants was crucial in determining the success and survival of some of the early European explorers.

The paintings of Albert Namatjira and other Aboriginal artists have made the beautiful Ghost Gums famous.

Another well-known Gum Tree is the Coolabah which figures prominently in Australian folklore and songs. For instance, Burke, Wills, and King—great explorers—found the historic "Dig" message on a Coolabah near Cooper's Creek in south-western Queensland. It has been preserved to this day, and a stone cairn nearby commemorates the ill-fated expedition.

Most of Australia's honey comes from the flowers of the Eucalypts. Some species provide tannin, and others valuable timber.

The Mountain Ash, one of the tallest hardwoods in the world, is also the world's tallest flowering tree. In the Cumberland Valley, Victoria, there are Mountain Ash growing to more than 90·36 metres (300 feet) high, but Tasmania has the tallest known measuring 99·96 metres (322 feet).

The koala bear, one of Australia's marsupials, lives entirely on Gum leaves but it is very selective as to the species.

The dye colours from the Eucalypts are very beautiful. Ranging from red, through various shades of orange to yellow and olive-green, they fittingly complement our "sun-burnt country."

EUCALYPTUS ALBA—Poplar Gum or White Gum

LEAVES

Mordant

alum—yellow
tin—yellow
copper—light brown
iron—grey
iron and copper—dark grey
iron and tin—grey

EUCALYPTUS AMYGDALINA — Black Peppermint

LEAVES

Mordant

alum—lemon
tin—gold
copper—light brown
iron—grey
iron and copper—dark grey
iron and tin—brown-gold

EUCALYPTUS AROMAPHLOIA—Scent-bark

LEAVES

Mordant

alum—orange

EUCALYPTUS BAXTERI—Brown Stringybark

LEAVES

Mordant

alum—green and yellow-fawn
tin—yellow
copper—brown
iron—grey
iron and copper—dark grey

EUCALYPTUS BEHRIANA—Bull Mallee or Broad-leaved Mallee Box

LEAVES

Mordant

alum—green-gold

EUCALYPTUS BICOSTATA—Victorian Blue Gum or Southern Blue Gum

LEAVES

Mordant

alum—orange-tan

EUCALYPTUS BLAKELYI—Blakely's Red Gum

LEAVES

Mordant

alum—fawn
tin—dark gold
copper—brown
iron—grey-brown
iron and copper—dark grey
iron and tin—green-gold

EUCALYPTUS BOSISTOANA—Bosisto's Box, Coast Grey Box, or Gippsland Grey Box

LEAVES

Mordant

alum—tan

EUCALYPTUS BOTRYOIDES—Bangalay, Southern Mahogany, or Mahogany Gum

LEAVES

Mordant

alum—lemon, yellow, and fawn
copper—green
chrome—brown-tan
alum, iron, and copper—green
alum and acetic acid—yellow
(no mordant—tan and green-gold)

LEAVES AND STEMS WITH SCALE INSECTS

Mordant

alum—pink-fawn

BARK

Mordant

alum—fawn-brown

One of Australia's best-known trees: the Coolabah Tree at Cooper's Creek on which the message "Dig" was left for the explorers Burke, Wills, and King

EUCALYPTUS BREVIFOLIA — Snappy Gum

LEAVES

Mordant

alum—orange-red

EUCALYPTUS BRIDGESIANA (syn. E. STUARTIANA)—Apple Box, But-But, Woollybutt, or White Peppermint

LEAVES

Mordant

alum—red

EUCALYPTUS CALOPHYLLA—Marri

LEAVES

Mordant

alum—yellow
alum and cream of tartar—dull gold

EUCALYPTUS CALYCOGONA — Red Mallee or Gooseberry Mallee

LEAVES

Mordant

alum—yellow-orange

EUCALYPTUS CAMALDULENSIS (syn. E. ROSTRATA)—River Red Gum, Murray Red Gum, Red Gum, or Queensland Blue Gum

LEAVES

Mordant

alum—lemon, yellow, orange-tan, tan, and pale tan
tin—tan, gold, and lemon.
copper—green, dull green, light brown, and brown
iron—dark brown, grey, and pale grey
iron and copper—dark brown and dark grey
alum and iron—brown
iron and tin—green

EUCALYPTUS CAMPHORA — Broad-leaved Sally or Swamp Gum

LEAVES

Mordant

chrome—dark gold

EUCALYPTUS CARNEA—White Mahogany or Yellow Stringybark

LEAVES

Mordant

alum—fawn-yellow

EUCALYPTUS CEPHALOCARPA—Silver-leaf Stringybark or Long-leaved Argyle Apple

LEAVES

Mordant

alum—dark red, red-tan, and orange
alum and tin—bright orange
alum, tin, and copper—brown
alum, tin, copper, and iron—brown
copper—green-fawn
copper and tin—gold
copper, tin, and iron—fawn-green
tin—yellow
tin and iron—gold
tin, iron, and copper—grey-green
tin, iron, copper, and alum—grey-green
iron—grey

FLOWERS

Mordant

alum—yellow-orange

BARK

Mordant

alum—fawn-gold
chrome—grey-fawn

FRUIT

Mordant

alum—fawn-yellow

EUCALYPTUS CINEREA—Mealy Stringybark, Blue-leaved Apple, Argyle Apple, or Blue Peppermint

LEAVES

Mordant

alum—red
chrome—brown-tan
(no mordant—pink-brown)

BARK

Mordant

alum—fawn-brown
chrome—brown
iron—brown

EUCALYPTUS CITRIODORA—Lemon-scented Gum

LEAVES

Mordant

alum—yellow and fawn-yellow
(no mordant—fawn-brown)

BARK

Mordant

alum and acetic acid—yellow

EUCALYPTUS CLADOCALYX — Sugar Gum

LEAVES

Mordant

alum—lemon and yellow
tin—yellow
copper—green
iron—grey
iron and copper—grey
chrome—yellow-tan
alum and cream of tartar—yellow-fawn and fawn
alum and copper—fawn-tan

EUCALYPTUS CONSIDENIANA—Yertchuk or Prickly Stringybark

LEAVES

Mordant

alum—yellow

EUCALYPTUS COOLABAH—Coolabah (Blue-leaf and Green-leaf)

LEAVES

Mordant

alum—lemon and yellow
tin—yellow and gold
copper—green-grey, fawn-green, and brown
iron—grey
iron and copper—dark grey
tin and iron—gold and dark gold

SAPWOOD CHIPS

Mordant

acetic acid—grey

EUCALYPTUS CORDATA—Heart-leaf Silver Gum or Tasmanian Silver Gum

LEAVES

Mordant

alum—red
copper—red-tan
tin—orange
iron—grey
iron and copper—grey-green
iron and tin—green

EUCALYPTUS CREBRA—Narrow-leaved Red Ironbark or Narrow-leaved Ironbark

LEAVES

Mordant

alum—fawn and dull yellow
tin—gold
copper—dull green
iron—grey
copper and iron—dark grey
tin and iron—dull gold

BARK AND KINO

Mordant

chrome—pink-brown

EUCALYPTUS CRENULATA—Silver Gum

LEAVES

Mordant

alum—yellow
iron and acetic acid—green
(no mordant—orange-tan)

EUCALYPTUS CYPELLOCARPA—Mountain Grey Gum

LEAVES

Mordant

alum—orange

EUCALYPTUS DALRYMPLEANA—Mountain Gum, Broad-leaved Ribbon Gum, or White Gum

LEAVES

Mordant

alum—red-orange

EUCALYPTUS DEALBATA — Tumble-down Gum or Hill Red Gum

LEAVES

Mordant

alum—fawn
tin—gold
copper—light brown
iron—grey-mauve

EUCALYPTUS DIVERSICOLOR—Karri

LEAVES

Mordant

alum—tan
tin—yellow
copper—brown
iron—grey

The leaves of the Ghost Gum produce shades of dye varying from grey to yellow-orange

EUCALYPTUS DIVES—Broad-leaf Peppermint or Blue Peppermint

LEAVES

Mordant

alum—yellow

EUCALYPTUS DUMOSA—White Mallee, Congoo Mallee, or Dumosa Mallee

LEAVES

Mordant

alum—yellow-orange

EUCALYPTUS FASCICULOSA — Pink Gum

LEAVES

Mordant

alum—tan
tin—bright yellow
copper—light brown
iron—grey
iron and copper—dark grey
iron and tin—green
iron and alum—grey-green

EUCALYPTUS FIBROSA—Broad-leaved Ironbark

LEAVES

Mordant

alum—yellow

EUCALYPTUS FICIFOLIA—Western Australian Red Flowering Gum

LEAVES

Mordant

alum—fawn and yellow
tin—yellow
copper—light brown
iron—black
(no mordant—fawn-brown)

BARK

Mordant

alum—fawn-brown

FRUIT

Mordant

alum—fawn-brown

EUCALYPTUS FOECUNDA—Narrow-leaf Red Mallee or Slender-leaved White Mallee

LEAVES

Mordant

alum—yellow

EUCALYPTUS FOELSCHEANA — Fan-leaved Bloodwood

LEAVES

Mordant

alum—yellow-brown
tin—yellow
copper—brown
iron—grey
iron and copper—grey
iron and tin—green

EUCALYPTUS GLOBOIDEA—White Stringybark

LEAVES

Mordant

alum—yellow

EUCALYPTUS GLOBULUS — Southern Blue Gum, Blue Gum, or Tasmanian Blue Gum

LEAVES

Mordant

alum—yellow and orange
iron—grey-fawn
(no mordant—gold-brown)

FRUIT

Mordant

alum—fawn
chrome—yellow-fawn

EUCALYPTUS GONIOCALYX—Bundy, Long-leaved Box, or Mountain Apple

LEAVES

Mordant

alum—lemon, yellow, and light tan
chrome—yellow-gold, fawn-green, and green-brown

EUCALYPTUS GRACILIS—White Mallee, Mallee Gum, or Yorrell

LEAVES

Mordant

alum—yellow

EUCALYPTUS GRANDIS—Flooded Gum or Rose Gum

LEAVES

Mordant

alum—tan
tin—yellow
copper—brown
iron—black

EUCALYPTUS GROSSA—Phillips River Gum or Coarse-leaved Mallee

LEAVES

Mordant

alum—lemon

EUCALYPTUS HUBERANA — Rough-barked Manna Gum

LEAVES

Mordant

alum—lemon
tin—yellow
copper—green
iron—grey
iron and copper—dark grey
iron and tin—gold green
tin, iron, copper, and alum—yellow-green
tin, iron, and copper—brown-green
alum, iron, and copper—green

EUCALYPTUS INCRASSATA—Lerp Mallee or Yellow Mallee

LEAVES

Mordant

alum—gold

EUCALYPTUS INTERMEDIA—Pink Bloodwood

LEAVES

Mordant

alum—yellow
tin—gold
copper—grey-green
iron—grey

EUCALYPTUS LARGIFLORENS—Black Box, River Box, or Flooded Box

LEAVES

Mordant

alum—yellow

EUCALYPTUS LEUCOXYLON—South Australian Blue Gum, White Ironbark, or Yellow Gum

LEAVES

Mordant

alum—lemon and yellow
tin—yellow
copper—light brown
iron—grey
iron and copper—dark grey

EUCALYPTUS LEUCOXYLON ROSEA—Pink-flowering Gum or Pink-flowered Yellow Gum

LEAVES

Mordant

alum—pale orange

EUCALYPTUS LINEARIS—White Peppermint or Peppermint Gum

LEAVES

Mordant

alum—lemon
chrome—green-lemon
iron—lemon
(no mordant—fawn)

EUCALYPTUS MACROCARPA—Rose-of-the-West, Mottlecah, or Desert Mallee

LEAVES

Mordant

alum—orange
tin—yellow
copper—fawn
iron—dark grey

EUCALYPTUS MACRORRHYNCHA—Red Stringybark

LEAVES

Mordant

alum—lemon
chrome—brown-green

EUCALYPTUS MACULATA—Spotted Gum

LEAVES

Mordant

alum—yellow

EUCALYPTUS MANNIFERA—Rough-barked Ribbon Gum, Manna Gum, Brittle Gum, or Cabbage Gum

LEAVES

Mordant

alum—fawn
tin—gold
copper—green-brown
iron—brown
copper and iron—dark brown
tin and iron—green-brown

The brilliant flowers of the Western Australian Red Flowering Gum. The bark, leaves, and fruit of this tree will yield dye colour

EUCALYPTUS MELANOPHLOIA — Broad-leaf Ironbark or Silver-leaf Ironbark

LEAVES

Mordant

alum—fawn
tin—gold
copper—fawn-green
iron—grey

EUCALYPTUS MELLIODORA—Yellow Box or Honey Box

LEAVES

Mordant

alum—brown, light tan, and gold
tin—bright yellow
copper—pale green
chrome—brown-tan

EUCALYPTUS MICRANTHA—Squiggly Gum

LEAVES

Mordant

alum—yellow

EUCALYPTUS MICROCARPA — Grey Box, Black Box, Inland Grey Box, or Green-leaved Box

LEAVES

Mordant

alum—lemon, yellow, and tan
tin—yellow
copper—fawn-green
iron—dark grey
iron and copper—black
iron and tin—green
chrome—brown-tan

EUCALYPTUS MICROCORYS—Tallow-wood

LEAVES

Mordant

alum—orange-tan

EUCALYPTUS MINIATA—Darwin Woollybutt or Melaleuca Gum

LEAVES

Mordant

alum—fawn
tin—yellow
copper—brown-green
iron—grey
iron and copper—dark grey
iron and tin—dull green

EUCALYPTUS MITCHELLIANA—Mitchell's Gum

LEAVES

Mordant

alum—gold

EUCALYPTUS MOLUCCANA (syn. E. HEMIPHLOIA)—Gum-topped Box or Grey Box

LEAVES

Mordant

alum—fawn
tin—gold
copper—brown
iron—brown-purple

EUCALYPTUS MUELLERANA—Yellow Stringybark

LEAVES

Mordant

alum—yellow

EUCALYPTUS NICHOLI—Narrow-leaved Black Peppermint or Willow Peppermint

LEAVES

Mordant

alum—yellow-orange and orange-tan

EUCALYPTUS NOVA-ANGLICA—New England Peppermint or Black Peppermint of New England

LEAVES

Mordant

alum—lemon

EUCALYPTUS NUBILA—Ironbark

LEAVES

Mordant

alum—yellow and fawn-yellow
tin—orange
copper—green-brown
iron—grey
iron and copper—dark grey
tin and iron—dark gold
tin, copper, and iron—grey-green
alum and iron—grey-green

EUCALYPTUS OBLIQUA—Messmate or Messmate Stringybark

LEAVES

Mordant

alum—lemon, yellow, and fawn-yellow
tin—yellow
copper—green
iron—grey and dark grey
chrome—brown
iron and copper—dark green and dark grey
iron and tin—green-brown and green

BARK

Mordant

alum—fawn-brown
chrome and iron—green-brown

EUCALYPTUS OLEOSA—Oil Mallee or Giant Mallee

LEAVES

Mordant

alum—yellow

EUCALYPTUS OVATA—Swamp Gum

LEAVES

Mordant

alum—orange

EUCALYPTUS PANDA—Narrow-leaved Ironbark of New South Wales

LEAVES

Mordant

alum—red-tan and tan

tin—bright yellow

EUCALYPTUS PAPUANA—Ghost Gum, Tropical Gum, or Drooping White Gum

LEAVES

Mordant

alum—fawn and yellow-orange

tin—yellow

copper—grey

iron—dull yellow

EUCALYPTUS PAUCIFLORA—Snow Gum, White Sally, or Cabbage Gum

LEAVES

Mordant

alum—yellow

EUCALYPTUS PERRINIANA — Round-leaf Snow Gum or Spinning Gum

LEAVES

Mordant

alum—lemon-yellow

EUCALYPTUS PILLIGAENSIS—Narrow-leaved Box or Mallee Box

LEAVES

Mordant

alum—fawn

tin—gold

copper—brown

iron—mauve-grey

iron and copper—dark grey

iron and tin—green

EUCALYPTUS PLANCHONIANA—Bastard Tallow-wood

LEAVES

Mordant

alum—lemon

tin—yellow

copper—green

iron—grey

copper and iron—black

EUCALYPTUS POLYANTHEMOS—Red Box

LEAVES

Mordant

alum—yellow-gold, gold, orange, red-tan, and tan

chrome—fawn-grey, brown-tan, and yellow-fawn

iron and cream of tartar—green-grey

(no mordant—gold-green)

BARK

Mordant

alum—fawn-brown

chrome—fawn-green

EUCALYPTUS POLYCARPA—Small-fruited Bloodwood or Pale Bloodwood

LEAVES

Mordant

alum—lemon

tin—yellow

iron—dark grey

EUCALYPTUS POPULNEA—Poplar Box or Bimble Box

LEAVES

Mordant

alum—lemon and pale tan

tin—lemon

copper—light brown and green

iron—grey

YOUNG SUCKER-LEAVES AND TWIGS

Mordant

alum and cream of tartar—grey-fawn

EUCALYPTUS RADIATA—River White Gum or Narrow-leaf Peppermint

LEAVES

Mordant

alum—yellow-gold

chrome—brown

EUCALYPTUS REDUNCA—Wandoo

WOOD CHIPS

Mordant

alum—fawn

chrome—fawn-brown

tin—fawn-pink

iron—fawn-green

EUCALYPTUS REGNANS—Mountain Ash or Australian Mountain Ash
LEAVES
Mordant
alum—lemon-yellow
chrome—brown-green

EUCALYPTUS RISDONI—Silver Peppermint
LEAVES
Mordant
alum—dark fawn
tin—gold
copper—brown-green
iron—grey
iron and copper—black
iron and tin—brown

EUCALYPTUS ROBUSTA—Swamp Mahogany
LEAVES
Mordant
alum—green-lemon

EUCALYPTUS RUBIDA—Candlebark or White Gum
LEAVES
Mordant
alum—yellow

EUCALYPTUS SALIGNA—Sydney Blue Gum
LEAVES
Mordant
alum—light tan
tin—yellow
copper—light brown
iron—brown
iron and copper—dark grey
iron and tin—green

EUCALYPTUS SALMONOPHLOIA—Salmon Gum
LEAVES
Mordant
alum—yellow and dull yellow
tin—orange and gold
copper—dull green
iron—grey

EUCALYPTUS SARGENTI—Swamp Mallet, Sargent's Mallet, or Salt River Mallet
LEAVES
Mordant
alum—yellow-gold

EUCALYPTUS SEEANA—Narrow-leaved Grey Gum or Stone Gum
LEAVES
Mordant
alum—lemon
tin—yellow
copper—brown
iron—grey

EUCALYPTUS SIDEROXYLON—Red Ironbark or Mugga
LEAVES
Mordant
alum—orange
BARK
Mordant
iron—grey-brown
tin—light tan
copper—fawn

EUCALYPTUS SIEBERI (syn. E. SIEBERIANA)—Silvertop, Black Ash, or Silvertop Ash
LEAVES
Mordant
alum—lemon

EUCALYPTUS STAIGERANA—Lemon-scented Ironbark
LEAVES
Mordant
alum—fawn
tin—yellow
copper—green
iron—grey
iron and copper—dark grey
iron and tin—green

EUCALYPTUS STELLULATA—Black Sally or Muzzlewood
LEAVES
Mordant
alum—pale fawn

EUCALYPTUS STRICKLANDI—Strickland's Gum
LEAVES
Mordant
alum—lemon

tin—yellow
copper—brown
iron—grey
iron and copper—dark grey
iron and tin—grey-green

EUCALYPTUS TERETICORNIS (syn. E. UMBELLATA)—Red Gum, Forest Red Gum, or Slaty Gum

LEAVES

Mordant
alum—lemon, yellow, gold, and deep gold
chrome and alum—green-gold
alum and iron—brown

TWIGS AND YOUNG LEAVES

Mordant
alum—green

EUCALYPTUS TERMINALIS—Bloodwood or Western Bloodwood

LEAVES

Mordant
alum—fawn
tin—gold
copper—light brown
iron—light grey
iron and copper—grey
iron and tin—green-grey

EUCALYPTUS TESSELLARIS—Moreton Bay Ash or Carbeen

LEAVES

Mordant
alum—yellow
tin—gold
copper—brown-green
iron—grey

EUCALYPTUS TETRAGONA — Mealy Gum, White-leaved Marlock, or Tallerack

LEAVES

Mordant
alum—red
tin—gold
iron—grey-brown

EUCALYPTUS TETRODONTA—Darwin Stringybark

LEAVES

Mordant
alum—yellow
tin—yellow
copper—dull green
iron—grey
iron and copper—dark grey
iron and tin—green

EUCALYPTUS VIMINALIS—Manna Gum, Ribbon Gum, or White Gum

LEAVES

Mordant
alum—yellow
tin—yellow
copper—brown-green
iron—grey
iron and copper—dark grey
iron and tin—brown-green
alum and copper—dull green
alum and bicarbonate of soda—lemon
(no mordant—tan)

EUCALYPTUS VIRIDIS — Green Mallee

LEAVES

Mordant
alum—yellow-gold

EUCALYPTUS WILKINSONIANA—Small-leaved Stringybark

LEAVES

Mordant
alum—fawn and pale green
tin—yellow and green
copper—brown and yellow

EUCALYPTUS WOOLLSIANA—Narrow-leaved Grey Box or Mallee Box

LEAVES

Mordant
alum—lemon
tin—fawn
iron—grey

EUCALYPTUS YOUMANI—Stringybark or Youman's Stringybark

LEAVES

Mordant
alum—dark fawn

LEAVES AND BARK

Mordant
alum—fawn-brown

This handwoven woollen fabric shows some of the beautiful warm colours that can be obtained from vegetable dyes

THE EUCALYPT EXPERIMENTS

At a seminar of the Handweavers and Spinners Guild of Victoria in September 1967, some casual remarks about two different colours obtained from dyeing handspun wool with Eucalypt leaves aroused Mrs Jean Carman's curiosity. She started on a comprehensive series of experiments the following year.

For her early tests she used leaves from several different species of Eucalypts growing at Croydon, Victoria, and she discovered very striking contrasts in the dyes produced. It became obvious that a considerable number of tests covering a wide range of Eucalypts would be necessary to arrive at any useful conclusion.

With the help of her family and friends, the Botany Department of the University of Melbourne, the Field Naturalists, and especially the officers of the Forests Commission of Victoria, Jean Carman tested more than fifty species of Eucalypts native to Victoria, in addition to species like the Western Australian Flowering Gum and Lemon-scented Gum which are extensively cultivated. Many other Victorian native trees were tested during the same year.

The exciting results which Jean Carman recorded became, in due course, the basis of this book.

To carry out each of these experiments, she cut up the adult leaves and boiled them in water in an enamel saucepan, with the lid on, for one hour. Then she strained the liquid, added the wool, which she had already mordanted with alum, and boiled it for another twenty minutes. Throughout the tests she followed the same procedure and checked all weights and measurements carefully. The wool used was Border Leicester Cross.

Colours obtained were fast to the usual washing of wool and fairly fast to light. They ranged from red, through varying shades of orange, to yellow and olive green.

Comparison of the dyes within the different groups of Eucalypts used in these tests showed many similarities.

In the Stringybark group, the predominate colour was olive green with the exception of Messmate *(E. obliqua)* which was bright yellow, while Mealy Stringybark *(E. cinerea)* and Silver-leaf Stringybark *(E. cephalocarpa)* were red. The Box group was fairly uniform with colours in the orange shades; the exception was Black Box *(E. largiflorens)* which gave yellow. This may have been caused by seasonal conditions because it was tested during the winter months. Apple Box *(E. bridgesiana)* gave the same red colour as *E. cephalocarpa* and *E. cinerea.*

The Mallees were mainly yellow, though Yellow Mallee *(E. incrassata),* Dumosa Mallee *(E. dumosa),* and Red Mallee *(E. calycogona)* were more orange in shade.

The Peppermint group was distinctly yellow. The Red Ironbark *(E. sideroxylon)* was orange.

The largest group of Eucalypts, the Gums, fell into two sections: those with colours in the range between yellow and olive-green and those which were distinctly orange, namely, Swamp Gum *(E. ovata),* River Red Gum *(E. camaldulensis),* Mountain Grey Gum *(E. cypellocarpa),* Mountain Gum *(E. dalrympleana),* and Victorian Blue Gum *(E. bicostata).*

The barks of the trees of all groups produced varying shades of brown.

Of the tests made with dyes from Eucalypts of the same species growing in different areas of Victoria, most showed little or no colour variations. However, subsequent tests made with some species cultivated out of their natural climatic areas have shown wider variations.

The most intense shades were evident during the hot, dry, summer months of 1967, but less than twenty-four hours after the drought "broke," the colour of the Silver-leaf Stringybark *(E. cephalocarpa)* and Messmate *(E. obliqua)* changed dramatically, being many shades paler after the rain. During the wet winter months Mountain Ash *(E. regnans)* and the Snow Gums *(E. pauciflora* and *E. stellulata),* were a drab colour instead of a clear yellow. This colour change during the wet season could apply to all species. It may be that the moisture content of the leaves is greater then and so the percentage of dye material in a given weight of leaves will be less than that contained in the same weight of leaves during the dry season. Consistently good colours have been achieved by drying all Eucalypt leaves before preparing the dyebath.

Various mordants other than alum were also tried during the Victorian experiments. Chrome made very little colour variation while copper, tin, and iron gave a variety of shades and a whole range of colours was evident by mixing the mordants.

When Jean Carman left the temperate Victoria to live in sub-tropical Queensland her work with dyes from Eucalypts continued. Experiments with wool mordanted with alum produced disappointing results. With the exception of Tallow-wood *(E. microcorys),* which gave a light red, the only colour obtained was yellow.

Mordanting with copper, iron, and tin produced a great range of colours and shades—yellow, orange, red, green, brown, grey, and black. The black was a surprise achievement resulting from the use of a copper mordant with the addition of a little iron to the dyebaths made from leaves from the two Tallow-woods *(E. microcorys* and *E. planchoniana).* A deep-brown colour was also obtained from the bark of these trees.

With the assistance of forestry officers and interested friends in all the different states, Jean Carman was able to compare the dye colours from the same species of Eucalypts which are native to more than one state of Australia.

Messmate *(E. obliqua)* from South Australia, Victoria, Tasmania, and New South Wales.

Manna Gum *(E. viminalis)* from Victoria and Tasmania.

Sugar Gum *(E. cladocalyx)* from South Australia and Victoria. Although not really a native of Victoria it is grown extensively in that state.

Yellow Gum *(E. leucoxylon)* (or Blue Gum as it is known in South Australia) from Victoria, New South Wales, and South Australia.

Forest Red Gum *(E. tereticornis)* from Victoria, New South Wales, and Queensland, where it is usually known as Blue Gum.

Yellow Box *(E. melliodora)* from Victoria and New South Wales.

Grey Box *(E. microcarpa)* from South Australia and Victoria.

Poplar Box *(E. populnea)* from New South Wales and Queensland.

Although tests with the preceding list of Eucalypts showed no difference, there were variations in the dyes from the River Red Gum *(E. camaldulensis)* from South Australia, Victoria, New South Wales, Queensland, and the Northern Territory. The colours ranged from yellow to light red. The brightest came from a tree growing in the dry bed of the Todd River at Alice Springs.

The Western Australian Flowering Gum *(E. ficifolia),* grown in Victoria, gave a different shade from the same variety grown in its native state. There was also a noticeable difference in colour from the Salmon Gum *(E. salmonophloia)* grown in Western Australia and in Queensland.

LILLY-PILLY

EUGENIA SMITHI (syn. ACMENA SMITHI)—Lilly-pilly

LEAVES AND STEMS

Mordant

copper—green-fawn

(no mordant—pale green)

EUGENIA VENTENATI (syn. SYZYGIUM VENTENATI)—Weeping Lilly-pilly or Weeping Myrtle

LEAVES AND STEMS

Mordant

alum—green

copper—dark green

(no mordant—fawn-green)

NATIVE CHERRY

The Native Cherry has cypress-like foliage and is partly parasitic on the roots of other trees. At first the seeds of this tree were thought to grow outside the fruits, but it has been found that the edible fleshy part was really the stalk. The fruit or seed at the end of the stalk is round, hard, and only about 0·63 cm ($\frac{1}{4}$ inch) long. The "fruit" stalks are succulent in winter and spring. In the early days the settlers considered that the land on which the Native Cherry grew was suitable for apple culture.

EXOCARPUS CUPRESSIFORMIS—Native Cherry, Wild Cherry, or Cherry Ballart

LEAVES AND STEMS

Mordant

alum—fawn-yellow, yellow, and yellow-green

copper—gold-green and green

tin—orange

iron—green-grey

chrome—gold

FICUS MACROPHYLLA—Moreton Bay Fig

RIPE FRUIT

Mordant

alum—green-grey

chrome—green-fawn

FLINDERSIA MACULOSA—Leopard Wood

LEAVES

Mordant

alum—deep yellow

chrome—gold

tin and cream of tartar—yellow

copper—green

iron—grey

GEIJERA PARVIFLORA—Wilga

LEAVES

Mordant

alum—dull yellow

tin—lemon

copper—green

iron—green

GOODENIA OVATA—Hop Goodenia

LEAVES AND STEMS

Mordant

alum—lemon-yellow and yellow-gold

GREVILLEA

The Grevillea genus ranges from tall trees to ground-creeping shrubs. The flowers may be red, pink, orange, or white. The Silky Oak, the largest Grevillea, has attractive fern-like foliage, and the fiery orange flowers which appear in early summer yield an abundance of palatable honey. It is very widely cultivated although native only to New South Wales and Queensland.

GREVILLEA ROBUSTA—Silky Oak
LEAVES
Mordant
alum—orange-yellow, lemon, and fawn
chrome—fawn and gold
copper—yellow-fawn
iron—grey

GREVILLEA ROSMARINIFOLIA—Rosemary Grevillea
FLOWERS
Mordant
alum—fawn-gold
FLOWERS AND LEAVES
Mordant
alum—lemon-yellow

PINCUSHION HAKEA

This Hakea is a most attractive plant from Western Australia. The red blossoms with cream stigmas, projecting for some distance from the flowers, have earned it the popular name of "Pincushion."

HAKEA LAURINA—Pincushion Hakea
LEAVES
Mordant
alum—fawn
alum and tin—lemon
DRIED LEAVES
Mordant
alum—grey-fawn

PURPLE CORAL PEA

The brilliant purple flowers of this tough-stemmed, twining plant make it a conspicuous flower of early spring. It is widespread in Victoria, New South Wales, and south-eastern Queensland.

HARDENBERGIA VIOLACEA—Purple Coral Pea or Sarsaparilla
FLOWERS
Mordant
alum—grey-blue

HETERODENDRUM OLEIFOLIUM—Boonaree, Boonery, Bullock Bush, or Rosewood
LEAVES
Mordant
alum—fawn-tan
chrome—tan-orange
iron—grey

HIBISCUS

Hibiscus is the Greco-Latin name for a type of mallow, while the specific name commemorates Baron F. von Huegel who collected specimens during the early part of the nineteenth century. The Lilac Hibiscus grows to about 2·43 m (8 feet) and bears large lilac flowers from late winter to summer.

HIBISCUS HUEGELI—Lilac Hibiscus
FLOWERS
Mordant
cream of tartar—pink-fawn
chrome—green and pale lemon

HIBISCUS TILIACEUS—Cotton Tree or Cottonwood
LEAVES
Mordant
alum—fawn

tin—brown
copper—gold
iron—grey-green
iron and copper—dark grey-green

LAGUNARIA PATERSONI—Pyramid Tree, Cow-itch Tree, or Norfolk Island Hibiscus

LEAVES, STEMS, AND YOUNG SEED PODS
Mordant
alum—yellow
(no mordant—lemon-fawn)

TEA TREE

This is a small tree or shrub with greyish-green leaves and white flowers, found throughout Australia. The timber is used for fences, trellises, and stakes. It makes a very useful wind-break and sand stay.

The leaves of these trees are not used for making tea as the name suggests. Actually the name arose because one species, *Leptospermum scorparium,* was used to make a tea infusion as a remedy for scurvy during one of Captain Cook's expeditions.

LEPTOSPERMUM LAEVIGATUM—Coastal Tea Tree

LEAVES AND STEMS
Mordant
alum—yellow
copper—brown-green
(no mordant—fawn)

LEAVES
Mordant
alum—lemon-yellow

LEPTOSPERMUM MYRSINOIDES—Silky Tea Tree

LEAVES AND STEMS
Mordant
alum—lemon

LEPTOSPERMUM PHILICOIDES (syn. L. ERICOIDES)—Burgan

LEAVES AND STEMS
Mordant
alum—yellow

As its name suggests, the Coastal Tea Tree is mostly found growing on sand dunes by the seacoast

HONEY-MYRTLES AND PAPERBARKS

The flowers are like small bottlebrushes, and their colour varies from pink and mauve to yellow-white. The Melaleucas share the common name of Bottlebrush, and this causes confusion with the Callistemons.

Melaleuca linariifolia, one of the Paperbarks, has the common name of Snow-in-Summer, derived from the white flowers which cover the bush in early summer.

MELALEUCA ARMILLARIS—Bracelet Honey-Myrtle
LEAVES AND STEMS
Mordant
alum—lemon-fawn

MELALEUCA DECUSSATA—Cross-leaf Honey-Myrtle
LEAVES AND STEMS
Mordant
alum—lemon-fawn
(no mordant—lemon-fawn)

MELALEUCA ERICIFOLIA—Swamp Paperbark
LEAVES AND STEMS
Mordant
alum—yellow
chrome—green
(no mordant—lemon-fawn)

MELALEUCA HYPERICIFOLIA—Red-flowering Paperbark
LEAVES AND STEMS
Mordant
alum—lemon and yellow
(no mordant—fawn-yellow)

MELALEUCA INCANA—Grey Weeping Melaleuca, Sliver-grey Myrtle, or Grey Honey-Myrtle
LEAVES AND STEMS
Mordant
alum—yellow-green
(no mordant—fawn)

MELALEUCA LINARIIFOLIA—Flax-leaf Paperbark or Snow-in-Summer
LEAVES, STEMS, AND BARK
Mordant
alum—green-fawn

MELALEUCA NESOPHILA—Showy Honey-Myrtle
LEAVES, STEMS, AND SEED PODS
Mordant
alum—lemon-fawn

MELALEUCA QUINQUENERVIA—Broad-leaf Paperbark
LEAVES
Mordant
alum—fawn
tin—pale fawn
copper—light brown
iron—grey

MELALEUCA SQUARROSA—Scented Paperbark
LEAVES AND STEMS
Mordant
alum—lemon-fawn

MELALEUCA STYPHELIOIDES—Prickly Paperbark
LEAVES AND STEMS
Mordant
alum—lemon, lemon-fawn, and yellow
copper—green-fawn

The Prickly Paperbark is native to coastal New South Wales and southern Queensland, but is commonly cultivated in all states

MYOPORUM INSULARE—Boobialla

LEAVES AND STEMS

Mordant

alum—fawn-tan

copper—green

(no mordant—fawn)

OLEARIA ARGOPHYLLA—Musk Daisybush

LEAVES

Mordant

alum—yellow

OWENIA ACIDULA—Emu Apple

LEAVES

Mordant

alum—dull yellow

chrome—gold-tan

copper—fawn-green

iron—brown

alum and copper—green

PETALOSTIGMA PUBESCENS—Quinine Berry

LEAVES

Mordant

alum—lemon-yellow

chrome—yellow-gold

copper—dark gold

iron—grey-brown

RICE FLOWER

The Rice Flower has very tough bark, which the early settlers frequently used for tying up their parcels and bunches of flowers. In emergencies they used it for their boot laces. The Aborigines were known to use the bark as a band for fastening their belongings together and as material for making their bags and nets.

In Western Australia it is known as the Bread Flower.

Its botanical name, *Pimelea,* is derived from the Greek *pimele,* meaning fat, relating to the oily seeds. It belongs to the Daphne family.

PIMELEA FERRUGINEA—Pink Rice Flower

LEAVES AND STEMS

Mordant

alum—yellow

PITTOSPORUM

The Sweet Pittosporum or Australian Mock Orange has heavily-scented, small, white flowers, dark green, glossy foliage, and bright orange fruit. A native of the eastern states, it is extensively cultivated both for hedges and as a specimen tree.

The Banyalla, or Cheesewood, is a small tree approximately between 3·03 m and 9·10 m (10 and 30 feet) in height. It is generally found in gullies and sub-alpine areas of New South Wales, Victoria, and Tasmania. The flowers are yellow, touched with purple, and the fruits are attractive, rounded capsules. It is also grown as a specimen tree.

The generic name means "pitch-seed," referring to the glutinous, sticky nature of the substance around the seed.

PITTOSPORUM BICOLOR—Banyalla or Cheesewood

LEAVES

Mordant

alum—lemon

Left: *The green leaves of the Staghorn Fern.* Right: *The coastal-growing Sea-berry Saltbush has dark reddish berries*

PITTOSPORUM UNDULATUM—Sweet Pittosporum, Australian Mock Orange, or Native Daphne

LEAVES

Mordant

alum—lemon, yellow, and gold

BERRIES

Mordant

alum—yellow and pale lemon

PLATYCERIUM GRANDE—Staghorn Fern

GREEN LEAVES

Mordant

alum—pale tan

(no mordant—pink-fawn)

POMADERRIS ASPERA—Hazel Pomaderris

LEAVES

Mordant

alum—yellow

POMADERRIS LANIGERA—Woolly Pomaderris

LEAVES AND FLOWERS

Mordant

copper—green

PROSTANTHERA LASIANTHOS—Victorian Christmas Bush

LEAVES

Mordant

alum—gold

ROUND-LEAF MINT-BUSH

Masses of bright lilac flowers cover the bush in spring. When crushed the leaves smell like mint. The bush is widespread in Victoria and also grows in New South Wales, South Australia, and Tasmania.

PROSTANTHERA ROTUNDIFOLIA—Mint-bush or Round-leaf Mint-bush

LEAVES

Mordant

alum—lemon

copper and iron—green-grey

BRACKEN FERN

Native to all states, Austral Bracken is the most extensive Australian species of this cosmopolitan Fern. The young roots are rich in starch, and the Aborigines used them for food.

PTERIDIUM ESCULENTUM — Austral Bracken Fern

YOUNG FRONDS

Mordant

alum—green

ROOTS

Mordant

alum—fawn-tan

RHAGODIA BACCATA—Sea-berry Saltbush

BERRIES

Mordant

alum—pale orange

BERRIES, STEMS, AND FEW LEAVES

Mordant

alum—lemon

LEAVES AND STEMS

Mordant

alum—yellow

QUANDONG

The Quandong is a graceful small tree with greyish leaves and small greenish-coloured flowers followed by succulent edible bright red fruits. In the early days, when the Mallee district in north-western Victoria was newly settled and cultivated fruits were scarce, Quandong jam and jelly were quite popular. The Aborigines greatly relished the fruits and used the seeds or nuts for making necklaces. The oily kernels are also edible.

SANTALUM ACUMINATUM—Quandong or Native Peach

LEAVES

Mordant

alum—fawn

tin—tan-yellow

oxalic acid and tannic acid—fawn-tan

FRUIT

Mordant

alum—red-tan

alum and copper—fawn-tan

SANTALUM LANCEOLATUM—Sandalwood

LEAVES

Mordant

alum—gold

tin—tan

copper—green-brown

iron—grey-green

SENECIO VAGUS—Saw Groundsel

FLOWERS

Mordant

chrome—bright gold

KANGAROO APPLE

Usually the older leaves of this shrub are shaped like a kangaroo's paw. The flowers are deep blue, while the fruit is large and ranges from yellow-orange to red in colour. It is native to all the eastern states as well as to New Zealand, where it is called Poroporo.

SOLANUM AVICULARE—Kangaroo Apple

LEAVES AND STEMS

Mordant

alum—green-yellow

SYNCARPIA GLOMULIFERA—Turpentine Tree

LEAVES

Mordant

alum—orange-fawn

alum and iron—grey

Above: *The leaves, fruit, and flowers of the Kangaroo Apple.* Below: *The tiny, daisy-shaped flowers of the Saw Groundsel, when mordanted with chrome, give a bright gold dye*

WARATAH

The Waratah, native to the Gippsland area in Victoria, is a tree with smooth, dark leaves. In summer the crimson flowers are arranged in dense globular heads. Light brown in colour, with a beautiful grain, the timber resembles the American Sycamore and was used for furniture and tool handles.

A small genus of three species confined to south-eastern Australia, it includes the shrubby New South Wales Waratah, well known as the floral emblem of that state.

Telopea is Greek for "seen at a great distance." Waratah, the common name, is a word of Aboriginal origin, meaning "seen from afar."

TELOPEA OREADES—Victorian Waratah or Gippsland Waratah

LEAVES AND FLOWERS

Mordant

alum—pale orange

TETRATHECA CILIATA—Pink-eye or Tetratheca

FLOWERS

Mordant

alum—brown-fawn

THRYPTOMENE

Thryptomene is an aromatic bushy shrub with small narrow leaves. Very small white flowers cover it in spring. It is attractive and widely cultivated.

THRYPTOMENE CALYCINA—Thryptomene or Grampians Thryptomene

LEAVES AND STEMS

Mordant

alum—orange-gold and tan

TRISTANIA

Tristania, an evergreen, can exist on a limited supply of water and is widely planted as a street tree. It grows to a height of approximately 15 m (50 feet) in its natural rain forests along the coast of New South Wales and Queensland.

TRISTANIA CONFERTA—Brisbane Box, Brush Box, or Queensland Box

LEAVES

Mordant

alum—yellow-gold and lemon
tin—yellow
copper—light brown
iron—grey
iron and copper—dark grey
iron and tin—tan

BARK

Mordant

alum—fawn

SEED PODS

Mordant

alum—yellow-gold

VENTILAGO VIMINALIS—Supplejack or Vine Tree

LEAVES

Mordant

alum—dull yellow and yellow
tin—yellow
copper—green-brown
iron—grey-green
chrome—tan

GRASS TREE

The Xanthorroea species is uniquely Australian and belongs to the Liliaceous genera, as do the Lily and Iris. The long-living, slow-growing Grass Tree usually indicates either poor soil, land that is wet in winter and dry in summer, or land requiring drainage and liming. The long spikes of creamy-white flowers stand high above the grass-like leaves. Almost invariably the flowers on the north side of the spike are the first to open; thus it is possible, from observing the flowering spikes, to determine the points of the compass. These plants flower vigorously after bushfires.

The Aborigines extracted honey from the flowers by immersing them in water. They also ate the tender white base of the leaves, and gathered the very strong resin which these plants contain.

Present-day apiarists dislike the resin because the bees, plastering it over the insides of the hives, fill in all the crevices and cracks. Quite frequently the bees seal on the lid of the hive and almost close up the entrance.

XANTHORRHOEA MINOR—Small Grass Tree

YOUNG FLOWER-HEAD AND STEM

Mordant

alum—dull gold

MISTLETOE

About 200 species belong to the Loranthus genus. They are parasitic, shrubby plants which are very common on trees. Although they usually possess green leaves, most of them have no rooted attachment to the soil, and they derive water and mineral salts at the expense of the host plants. The berries of the Mistletoe are used in the preparation of bird-lime.

Many members of the Handweavers and Spinners Guild have used the Mistletoe as a dye plant but have not identified the species used.

The colours obtained were:

Mordant

alum—lemon, yellow, and deep yellow

copper—brown-green

(no mordant—green-lemon)

REEDS AND GRASSES

Although there is no indication that the Aborigines used plants for dyeing, they spun many of the reeds and rushes which grew along the riverbanks. They made bags from spun threads, which they arranged in loops similar to "Sprang" weaving. Fishing nets were knotted from the spun fibres of the reeds.

GAHNIA RADULA—Sword Grass or Thatch Saw-sedge

BLACK SEED HEADS

Mordant

alum—fawn-brown

LOMANDRA LONGIFOLIA—Long Mat-rush, Spiny-headed Mat-rush, or Sagg

SEED HEADS

Mordant

alum—yellow

iron—grey-green

chrome—fawn

LICHENS

Lichens are found encrusting rocks, on damp fences, tree trunks, and roof tiles or slates. Some are flat, others leafy. The colours vary from grey, green, brown, and black to orange.

Many members of the Guild experimented with various lichens which have not been identified; thus no specific names are included. The colours are all beautiful and intense except that from the lichen commonly known as Old Man's Beard. This gave rather disappointing results. Sometimes a small amount of acetic acid put into the lichen dyebath seems to assist in releasing the colour.

Lichens from Australia, like those from the rest of the world, need no mordant because they are substantive or contact dyes, although some dyers use small quantities of mordants to add depth of colour.

Many colours are obtained—fawn, orange, brown, lemon, red-brown, tan, orange-tan, and very dark brown.

Lichens are contact dyes, and need no mordant

SELECT REFERENCE WORKS

Anderson, R. H. *The Trees of New South Wales.* Sydney: New South Wales Government Printer, 1968. Baglin, D. and Mullins, B. *Australian Wattles.* Sydney: Horwitz Publishing Inc., 1970. Black, J. M. *Flora of South Australia.* Parts I and II. Adelaide: South Australian Government Printer, 1922–29. Blake, S. T. and Roff, C. *The Honey Flora of South-East Queensland.* Brisbane: Queensland Department of Agriculture and Stock, 1958. Blakely, W. F. *A Key to the Eucalypts.* Canberra: Forestry and Timber Bureau, 1965. Brookes, A. E. *Tree Wonders of Australia.* Melbourne: Heinemann, 1965. Cochrane, G. R., Furher, B. A., Rotheram, E. R., Willis, J. H. *Flowers and Plants of Victoria.* Wellington: A. H. & A. W. Reed, 1968. Costerman, L. F. *Trees of Victoria.* Melbourne: Renwick Pride, 1966. *Dye Plants and Dyeing.* New York: Brooklyn Botanical Gardens, 1964. Galbraith, J. *Wildflowers of Victoria.* Melbourne: Longmans, 1967. Gardner, C. A. *West Australian Wildflowers.* Perth: West Australian Newspapers Ltd, 1945. Harris, T. Y. *Wildflowers of Australia.* Sydney: Angus & Robertson, 1966. Holliday, I and Hill, R. *A Field Guide to Australian Trees.* Adelaide: Rigby Ltd, 1969. *Introduction to the Victorian Forest Trees.* Melbourne: Forests Commission of Victoria, 1961. Kelly, S. *Eucalypts.* Melbourne: Thomas Nelson, 1969. Lloyd, J. *Dyes from Plants of Australia and New Zealand.* Wellington: A. H. & A. W. Reed, 1971. Millett, M. *Australian Eucalypts.* Melbourne: Lansdowne, 1969. Penfold, A. R. and Willis, J. L. *The Eucalypts.* London: Leonard Hill (Books) Ltd, 1961. Thurstan, V. *The Use of Vegetable Dyes.* Leicester: Dryad Press, 1942.

ACKNOWLEDGMENTS

Thanks are due to the members of the Handweavers and Spinners Guild of Victoria and all dyers who contributed the findings of their experiments; Mr and Mrs J. McL. Carman, St Lucia, Queensland; Miss K. Carman, Department of Agriculture, Stock, and Fisheries, Rabaul, Papua New Guinea; Mrs Jess Brooke, Glen Iris, Victoria; Mr L. Jacobson, Gardenvale, Victoria; Dr D. H. Ashton, Botany Department, University of Melbourne; Mr A Threader, Forests Commission, Victoria; Mr J. Anderson, Chemistry Department, Royal Melbourne Institute of Technology; Mr F. Rogers, Society for Growing Australian Plants; Professor and Mrs D. Carr, Botany Department, Research School of Biological Studies, Australian National University, Canberra; Dr H. T. Clifford, Botany Department, University of Queensland; the officers of the Forestry Department in each State; and the curators of the Herbariums of Victoria and Queensland.

For the photographs, the authors are grateful to Jocelyn Burt, Pauline Reynolds, Garth Marks, Jeanne Lamprell, Shirley Lucas, Val Marks, and Valda Martin.

INDEX TO COMMON NAMES

INDEX TO BOTANICAL NAMES